THE LIBRARY OF FOOD CHAINS AND FOOD WEBS

Food Chains in a

TIDE POOL HABITAT

ISAAC NADEAU
Photographs by
DWIGHT KUHN

The Rosen Publishing Group's
PowerKids Press
New York

For Grammy, and her love of life—Isaac Nadeau
For Savanah—Dwight Kuhn

Published in 2002 by The Rosen Publishing Group, Inc.
29 East 21st Street, New York, NY 10010

First Edition

Book Design: Emily Muschinske
Project Editor: Emily Raabe

Photo Credits: All Photos © Dwight Kuhn except pp. 8 (plankton), 9 (diatoms) , 10, and 20 (plankon and diatoms).

Nadeau, Isaac.
Food chains in a tide pool habitat / Isaac Nadeau.
     p. cm. — (The library of food chains and food webs)
Includes bibliographical references (p.   ).
  ISBN 0–8239–5761–6 (lib. bdg.)
1. Tide pool ecology—Juvenile literature. 2. Food chains (Ecology)—Juvenile literature. [1. Tide pool ecology. 2. Tide pools. 3. Food chains (Ecology) 4. Ecology.] I. Title. II. Series.
  QH541.5.S35 N34 2002
  577.69'9—dc21
                                                                                  00–013021

Manufactured in the United States of America

Contents

Food Chains in a Tide Pool

A crab could not hunt without energy. A barnacle could not reproduce without energy. Seaweed could not grow without energy. Kids could not explore without energy. Without energy, there would be no food chains in a tide pool. There would be no life at all!

Where can you find a creature with 30 eyes and one foot? Where can you find an animal that eats with its stomach outside its body? Where can you find a plant with no roots? All of these and many more amazing creatures are part of the food chains in a tide pool.

A food chain is a way of describing how plants and animals live and depend on each other for food. Each time an animal eats a plant or another animal, another link is formed in the food chain.

Each living thing gets energy from its food. Energy is where living things get the power to move about, to grow, and to do all of the important jobs they must do to stay alive.

This green algae, periwinkle, starfish, and herring gull form one of many food chains that you might find in a tide pool.

A Home Where the Land Meets the Sea

The plants and animals that live in a tide pool are very different from the plants and animals that live on land. That's because every living creature is **adapted** to its own **environment**. A place where a plant or an animal finds all that it needs to live is called a habitat. The tide pool is a perfect habitat for many different things. Mussels and oysters cling tightly to rocks in the tide pool. Crabs find shelter from the breaking waves. Eelgrass soaks up the sunlight with its underwater leaves. Periwinkles scrape the rock in search of food. All of the creatures that are part of the food chains in a tide pool habitat are adapted to catch their food among the waves. The creatures that live in a tide pool are adapted to a life of salt water and of waves pounding against the rocky shore.

Tide pool creatures are adapted to a world that is always in motion, as the waves roll in and out.

Riding the Tide

Tides are the rising and falling of water in the oceans. How do tides work? If you go to the ocean at night, you might see the Moon. The Moon is the cause of the tide. As the Moon **revolves** around Earth, its **gravity** pulls the tides up the beach. When the Moon passes, the tide slips back down. Tide pools are formed on rocky shores when the tide rises, bringing water high up on the shore. As the tide goes out, pockets of water are left behind in the thousands of holes, bowls, and cracks in the rock. These natural pools are the tide pool habitat. The world of the tide pool is always in motion, as the waves wash in and out on the tide. Sometimes the waves wash an unlucky plant or animal out to sea. Waves are important in a tide pool, though. They carry tiny bits of fresh food with them when they roll in. This food, called **plankton**, is made of **microscopic** plants and animals.

Right: *Microscopic plankton live in tide pools. Many of these animals are the tiny babies of the creatures that live in the tide pool. Plankton is a very important food source in a tide pool food chain.*

8

Sunshine for Breakfast

Some plants that live in the tide pool are not drifters. They hold themselves tightly to the rock, waving back and forth in the waves. One of the most common of these plants is the sea lettuce. It looks like true lettuce, but its leaves can be up to 4 feet (1.2 m) long.

The first link in a tide pool food chain is made up of plants. Plants produce all of the energy in the tide pool through **photosynthesis**. Photosynthesis is the ability of plants to take the energy from the Sun and combine it with water, air, and other **nutrients** to make sugar. This sugar is the plants' food. Plants use sugar to grow new leaves, to **reproduce**, and to make more sugar. Whatever energy is not used is stored in the plants' bodies. Animals that eat plants use this energy. One of the most important kinds of plants in the ocean is called **phytoplankton**. Phytoplankton are so small that we can't see them without a microscope.

Below: Red Algae has flat leaves. These leaves float near the surface to get sunlight. Right: These diatoms are phytoplankton that are covered in glasslike material.

Saltwater Salad

Limpets slowly scrape the **algae** off the rock. Sea urchins nibble at the base of seaweed. Shrimp sweep phytoplankton into their mouths. These are all members of the second link in a tide pool food chain, the **herbivores**. Herbivores are animals that eat plants. The energy the plant collected from the Sun is now passed on to the herbivores.

Most of the herbivores that live in a tide pool move very slowly. Often they have thick shells to protect them from the animals that want to eat them. Some kinds of sea slugs are herbivores. The sea slug is a soft, slow-moving creature that seems like it would be the perfect food for a hungry hunter. However, the sea slug tastes awful! Many of the herbivores in a tide pool food chain are types of snails. Some snails are herbivores, and some snails eat herbivores!

Above: *Limpets are herbivores that live in tide pools.*

Right: *Although the sea urchin is an herbivore, it sometimes eats other things.*

Below: *Sand dollars eat algae at the bottom of the tide pool.*

The Hunters and the Hunted

In a tide pool, it can be hard to tell who is the hunter and who is being hunted. Often an animal is hunting and being hunted at the same time! Animals that eat other animals are called **carnivores**. Carnivores are the third link in a tide pool food chain. Some carnivores, such as the octopus, move about in search of **prey**. Other carnivores, such as barnacles and anemones, sit and wait for their food to drift by and then catch it. Even carnivores are not safe in the tide pool, though. Many carnivores in the tide pool eat other carnivores. Carnivores that eat other carnivores are called secondary carnivores. Dog whelks and oyster drills are two kinds of secondary carnivores. They use their tongues to drill holes into the shells of their prey, and then they suck prey out of the holes. Secondary carnivores are the fourth link in a tide pool food chain.

Right: A starfish uses its powerful arms to pry open the shells of clams and mussels. Once the shell is open, the starfish passes its stomach through its mouth and into the shell to digest its prey.

Above: A sea anemone
waits for its food to
arrive.

Inset: These
barnacles are
using their "feet"
to strain food
from the
seawater.

Another common scavenger in the tide pool is the rock crab. Rock crabs eat the dead bodies of many different kinds of animals. They also eat algae, worms, and other things they might catch in their claws.

A Well-Balanced Diet

Almost nothing goes to waste in a tide pool food chain. Anything that still has a bit of energy stored in it is bound to end up in something else's stomach. Even dead plants and animals are gobbled up. Some of the most important parts of a tide pool food chain are the **scavengers**. Scavengers are animals that get their energy from the bodies of dead animals. A tiny insect called an anurida is a scavenger, and it feeds on dead crabs, mussels, barnacles, and fish. Hermit crabs are tide pool scavengers, too. Only the front of a hermit crab's body is protected with a shell. This means that hermit crabs must live in empty shells to protect their back ends.

Hermit crabs eat both plant matter and animal matter in the tide pool. Because so many animals eat other animals in a tide pool, there are often five or six links in a tide pool food chain!

Breaking It Down

Sponges are another kind of animal that lives in a tide pool habitat. Like rock crabs, sponges are scavengers, but they get their food in a very different way. Sponges get their food by sucking water in and out of their **pores**, which sweeps bits of dead animals, plants, and **bacteria** into their bodies.

Bacteria, the tiniest creatures in the ocean, are important to the food chain because they help to break down dead things into the smallest parts. They also are an important food for many other creatures, including sponges, mussels, clams, and oysters. All of the links in a tide pool food chain depend on bacteria to keep the energy that is stored in dead bodies from going to waste.

Right: *After bacteria break down the body of this dead crab* (inset), *it will become food for many small creatures in the tide pool, such as this sea sponge* (large photo).

The Invisible Web

Every member of a tide pool food chain is part of an even bigger food web. A food web shows how the health of one plant or animal depends on the health of all the others that share its habitat. The arrows in this food web point to the creatures that are getting energy. When the creatures in food webs die, such as the dead crab at lower left, decomposers will make sure that the energy from the creatures' bodies is used to begin still more tide pool food chains.

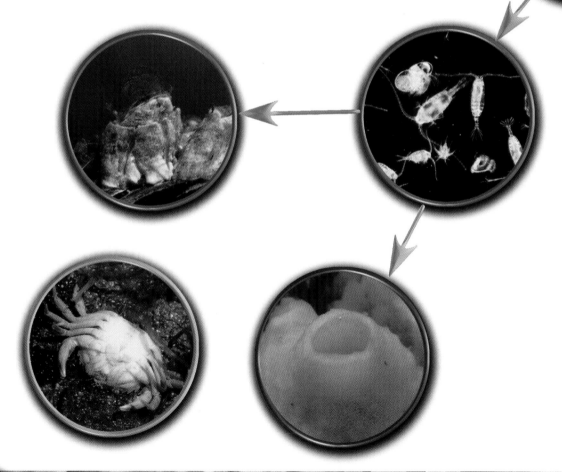

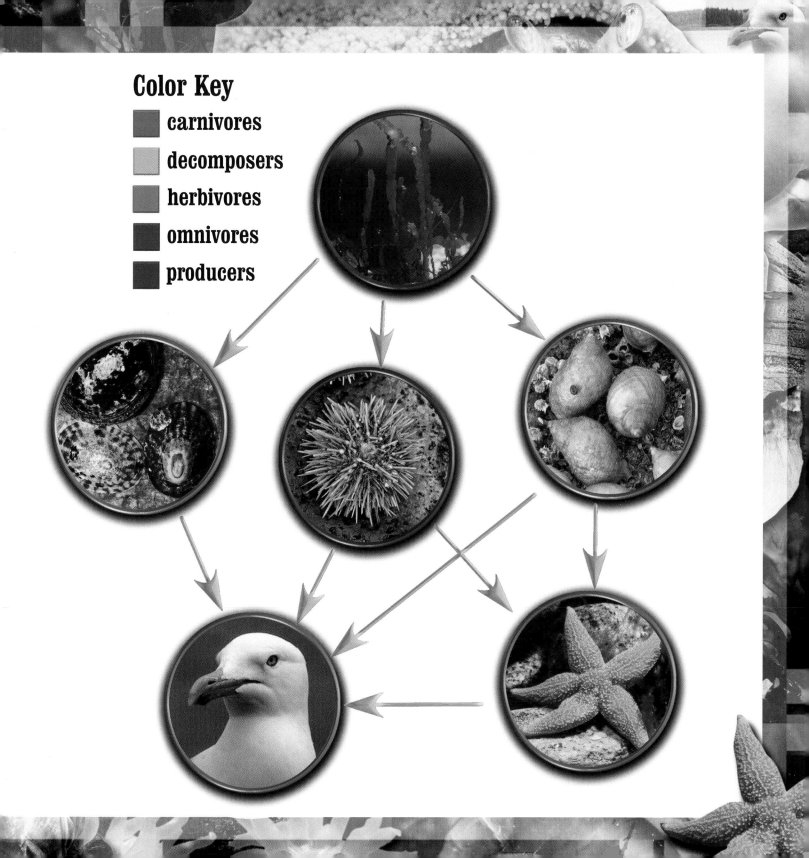

Color Key

- carnivores
- decomposers
- herbivores
- omnivores
- producers

Exploring a Tide Pool Habitat

Approach tide pools slowly and be patient. After all, the creatures that live there can't be sure that you are not a predator!

Low tide is the best time to explore a tide pool, because that's when the water is shallowest and the pools are easiest to reach. If you go, you'll need to wear shoes to protect your feet from sharp rocks and shells. Avoid stepping on fragile plants and animals like mussels, and be sure to put back anything you pick up. Try comparing two or three different pools to one another. Choose pools of different sizes and at different distances from the ocean. Do you see some plants and animals in one pool that are not in the others? Tide pools are filled with all kinds of amazing things. You can watch a tide pool all day and still not see everything that lives there.

Glossary

adapted (uh-DAPT-id) Changed to fit conditions.

algae (AL-jee) Plants without roots or stems that usually live in water.

bacteria (bak-TEER-ee-uh) Tiny living things that can only be seen with a microscope and that cause living things to decay.

carnivores (KAR-nih-vorz) Animals that eat other animals for food.

environment (en-VY-urn-ment) All of the living things and conditions that make up a place.

gravity (GRA-vih-tee) The natural force that causes objects to move or tend to move toward one another.

herbivores (ER-bih-vorz) Animals that eat plants for food.

microscopic (my-kroh-SKAH-pik) So small it can only be seen with a microscope.

nutrients (NOO-tree-intz) Anything that a living thing needs for its body to live and grow.

photosynthesis (foh-toh-SIN-thuh-sis) The process by which plants use energy from sunlight, gases from air, and water to make food and release oxygen.

phytoplankton (fy-toh-PLANK-ten) An ocean plant made up of one cell.

plankton (PLANK-ten) Plants and animals that drift with water currents. Many plankton are too small to see without a microscope.

pores (PORZ) Small holes.

prey (PRAY) An animal that is hunted by another animal for food.

reproduce (ree-pruh-DOOS) To make more of something of the same kind.

revolves (ree-VAHLVZ) Makes a circle around.

scavengers (SKA-ven-jurz) Animals that feed on dead animals.

Index

Web Sites

To learn more about tide pools, check out these Web sites:

www.mbayaq.org/efc/efc_hp/hp_rcky_cam.asp
www.mbayaq.org/efc/efc_hp/hp_rocky.asp
www.mbayaq.org/lc/kids_place/kidseq_equarium.asp